Shi Er Duan Jin

other books in the same series

Daoyin Yangsheng Gong Shi Er Fa
12-Movement Health Qigong for All Ages
Chinese Health Qigong Association
ISBN 978 1 84819 195 2

Da Wu
Health Qigong Da Wu Exercises
Chinese Health Qigong Association
ISBN 978 1 84819 192 1

Mawangdui Daoyin Shu
Qigong from the Mawangdui Silk Paintings
Chinese Health Qigong Association
ISBN 978 1 84819 193 8

Taiji Yangsheng Zhang
Taiji Stick Qigong
Chinese Health Qigong Association
ISBN 978 1 84819 194 5

Ba Duan Jin
Eight-Section Qigong Exercises
Chinese Health Qigong Association
ISBN 978 1 84819 005 4

Yi Jin Jing
Tendon-Muscle Strengthening Qigong Exercises
Chinese Health Qigong Association
ISBN 978 1 84819 008 5

CHINESE HEALTH QIGONG

Shi Er Duan Jin

12-Routine Sitting Exercises

CHINESE HEALTH QIGONG ASSOCIATION

SINGING DRAGON

LONDON AND PHILADELPHIA

This edition published in 2014
by Singing Dragon
an imprint of Jessica Kingsley Publishers
73 Collier Street
London N1 9BE, UK
and
400 Market Street, Suite 400
Philadelphia, PA 19106, USA

www.singingdragon.com

First published by Foreign Languages Press, Beijing, China, 2012

Library of Congress Cataloging in Publication Data
A CIP catalog record for this book is available from the Library of Congress

British Library Cataloguing in Publication Data
A CIP catalogue record for this book is available from the British Library

ISBN 978 1 84819 191 4

Printed and bound in China

CONTENTS

CHAPTER I

Origins and Development

Chapter I
Origins and Development

Shi Er Duan Jin is also known as a type of *Daoyin* in ancient China, which is a Chinese school of traditional body-building exercise combining breath control, body and limb movements, concentration of mind and local massage. It is made up of 12 routines. The word "*Jin*" means that as an integrated, sitting *Daoyin* exercise, the movements in these sequences look like a delicate, beautiful and continuous drawing.

The name *Shi Er Duan Jin* first appeared in the during the 18th century reign of Qianlong in the Qing Dynasty. The routine came from *Zhong Li Ba Duan Jin Fa* (钟离八段锦法, Zhongli Eight-routine Exercise) included in *Ten Works on Cultivating Perfection* (修真十书) compiled during the Ming Dynasty. Zhongli is the surname of Tang-Dynasty Taoist Zhongli Quan, founder of this exercise. The book gathered together dozens of important *Qigong* and *Neidan* works from the Sui, Tang and Song dynasties.

During the Ming Dynasty, *Zhong Li Ba Duan Jin Fa* was referenced in many health preservation works, so the method attracted great attention from health experts of the time.

Under the Jiajing reign (1521–1566) of the Ming, *Shi Liu Duan Jin* (十六段锦, 16-routine Exercise) was compiled on the basis of the sitting-posture *Ba Duan Jin* (八段锦, Eight-routine Exercise). It is characterized by each movement being designed to dispel a specific "pathogen." It "does not cure diseases but prevents diseases." Its unique function is to dispel "accumulated wind pathogens" and "pathogenic *Qi*" in the viscera and other parts of the body.

During the Qianlong reign (1736–1795) in the Qing Dynasty, Xu Wenbi added four further diagrams to *Zhong Li Ba Duan Jin Fa*, made some changes to its verse formulas and interpretation, and renamed it *Shi Er Duan Jin* (十二段锦, 12-routine Exercise). The original contents of *Zhong Li Ba Duan Jin Fa* were basically retained in the new version.

During the Xianfeng reign (1850–1861) of the Qing Dynasty, Pan Wei used Xu Wenbi's *Shi Er Duan Jin* as the central text, but supplemented it with *Fen Xing Wai Gong Jue* (分行外功诀). He referred to many medical works to further perfect *Shi Er Duan Jin* and included this exercise in his book *Important Arts for Preserving Health* (卫生要术) with the same verses and diagrams. *Fen Xing Wai Gong Jue* is organized into exercises for the heart, body, head, face, ear, eye, mouth, tongue, teeth, nose, hand, foot, shoulder, back, abdomen, waist, and kidneys.

Health Qigong, Shi Er Duan Jin was created and compiled by researching and editing *Zhong Li Ba Duan Jin Fa* and *Shi Er Duan Jin* according to the principles of *Qigong* and the physical and mental characteristics of people in modern society.

Shi Er Duan Jin has retained the essentials of both stillness and motion, as well as the emphasis on both the physical and mental practice in the original exercises. This collection has integrated such traditional *Qigong* methods as massage, *Daoyin,* meditation, and reflective contemplation.

CHAPTER II

Characteristics

Chapter II
Characteristics

1. "Coordinated Mind and Form" and "Combined Motion and Respiration"

"Coordinated mind and form" means that the mind is used to drive the body in order to coordinate form with mind during an exercise. "Mind" refers to mental activities during an exercise; any movement of the body requires the participation of the conscious or subconscious mind. During each exercise, mind concentration must track closely the characteristics and requirements of the movements, in order to effectively relax the body and mind, calm the emotions, eliminate distracting thoughts, directly arouse the circulation of *Qi* in the body, clear the appropriate meridian channels, control and prevent related disorders, and promote organ functions. *Shi Er Duan Jin* requires that the mind track the movements of the body. In other words, the mind should be concentrated on the standards, details and important positions of the movements. At the same time, this regimen also requires that the mind be focused on different body parts and their changes during the movements. Mental concentration should be moderate; excessive concentration can easily cause headaches, chest distress, and abdominal distention, and can interrupt the flow of blood and *Qi*. Therefore we should "stay between attachment and non-attachment" so as to balance body and mind.

"Combined motion and respiration" means the movements should be coordinated with the breath during each exercise. Movements should be synchronized with breath. In other words, movements should match the circulation of internal *Qi*. Gentle, slow, even and continuous movements will help us achieve gentle, even, deep and long breathing.

2. "Alternating Motion and Stillness", and "Simultaneous Cultivation of Form and Spirit"

"Alternating motion and stillness" primarily indicates that *Shi Er Duan Jin* includes both moving exercises and still exercises and is an organic combination of motion and stillness. In "simultaneous cultivation of both form and spirit," "form" means the body and "spirit" refers to mind. "Form is the container of spirit." Mind and body depend on and use each other.

Health preservation masters in all ages advocated mixing motion and stillness exercises and advise us not to overemphasize one at the expense of the other. In the traditional health preservation theories of China, great emphasis is placed upon "simultaneous cultivation of both interior and exterior" and "simultaneous cultivation of both form and spirit." While they advocate "nourishing the spirit and reducing consumption through stillness," they also advocate "nourishing form through motion but refraining from excessive motion." Practice will strengthen your limbs, calm the viscera, nourish the spirit, clear the meridians, and regulate blood and *Qi*.

3. Emphasis on Stretching and Massage

"Emphasis on stretching" means that we should fully stretch and direct our body using breath and mind during our exercise. Through a series of movements centered in the spinal column, including bending, stretching, rotating, folding the limbs and leaning, *Shi Er Duan Jin* strengthens our bones, muscles, joints, and ligaments throughout the body, lubricates our joints, softens tendons and ligaments, improves the flexibility and coordination of limbs, and improves our physical constitution.

Massage is an important element of traditional Chinese medicine. "Emphasis on massage" means that we should make a point of massaging specific parts of the body during exercise. By stimulating specific acupoints and channels, we can balance *yin* and *yang* and harmonize blood and *Qi* in the body.

CHAPTER III

Practice Tips

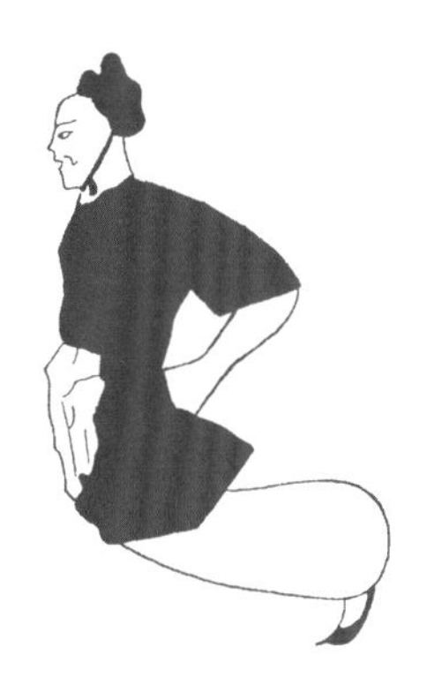

Section I Hand and Body Form

1. Basic Hand Shapes

The basic hand shapes of *Shi Er Duan Jin* include the natural palm, *Tongtian* finger, clenched fist, and square fist.

1) Natural Palm

Stretch the fingers out naturally and keep them apart (Fig. 1).

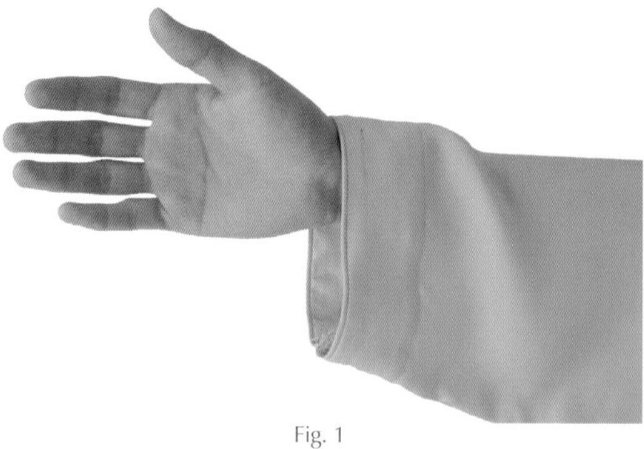

Fig. 1

2) Tongtian Finger

Stretch the fingers out naturally, slightly bending the middle finger downward and focus the mind on the fingertip of it (Fig. 2).

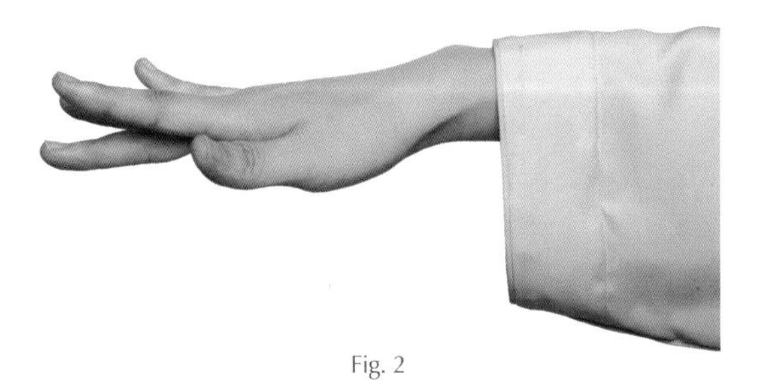

Fig. 2

3) Clenched Fist

Push the thumb up against the inner side of the third knuckle of the ring finger, clench the other four fingers and relax the *Laogong* acupoint (Fig. 3).

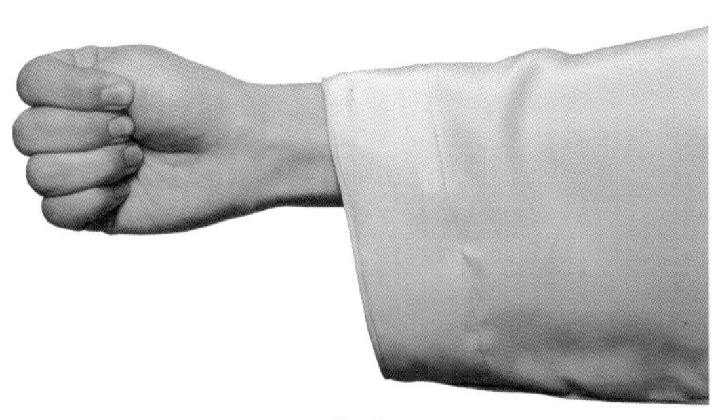

Fig. 3

4) Square Fist

Relax the four fingers naturally, with the thumb touching the second knuckles of the index and middle fingers. The surface of the upper part of the fist should be flat (Fig. 4).

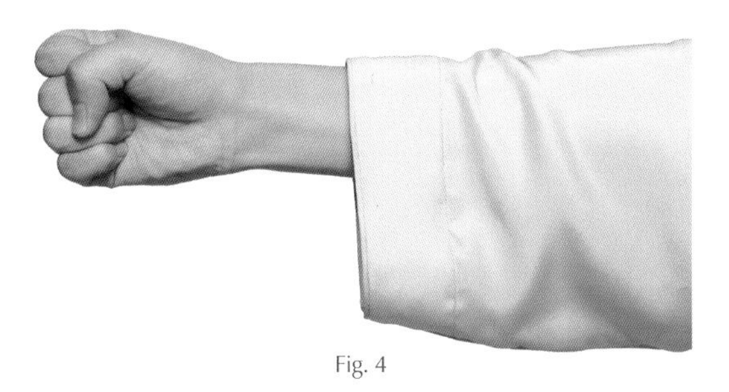

Fig. 4

2. Basic Body Form

Body attitudes are conscious adjustments of posture and movement, with the aim of meeting the form requirements and purposes of the exercises. Basic body postures of *Shi Er Duan Jin* include initial stillness and motion postures.

1) Stillness

Take Routine 10 "Burning Navel Ring" in the natural sitting posture as an example (Fig. 5):

Fig. 5

Sit erect, lift the *Baihui* acupoint, and pull in your chin. Lightly close your eyes, lips and teeth, and relax your expression, with the tongue touching the palate. Keep the neck and spine erect, relax your chest and abdomen, and keep your shoulders down and armpits slightly apart from body. Overlap your hands to gently massage the navel ring. Relax the buttocks and part your knees, turning the ankles to splay the thighs.

2) *Motion*

Crane the neck straight up, keep the neck and spine erect and relax the chest. Expand the shoulders and chest, relax the abdomen and buttocks, keep the shoulders and elbows down, and relax the wrists and fingers. The spine initiates the movements throughout the body. Keep your eyes on your hands, concentrate the mind, coordinate mind and form, and integrate motion with breathing.

Section II Breathing and Mind

1. Breathing

Breathing properly is an essential part of *Qigong* practice. Breath regulation means actively controlling your respiration to meet the specifications and accomplish the goals of these exercises. Beginners should breathe naturally and focus on learning the movements. After becoming familiar with the movements, regulate your breath according to the demands of the movements. With more experience, the exerciser's breathing will regulate itself unconsciously.

The specialized breathing forms of *Shi Er Duan Jin* include natural respiration, obverse abdominal respiration, reverse abdominal respiration, anus-lifting respiration, and holding the breath.

1) Natural Respiration

Natural respiration refers to breathing without conscious intervention of the mind. Beginners should start with this, which is indispensable in many exercises. Routines such as "Calming Heart and Positioning Hands," "Tapping Teeth and Sounding Drums" and "Rotating the Upper Body" in *Shi Er Duan Jin* all require natural breathing.

2) Obverse Abdominal Respiration

During obverse abdominal respiration, the lower abdomen expands during inhalation and retracts during exhalation. Routine 10 "Burning Navel Ring" requires this form of breathing.

3) Reverse Abdominal Respiration

During reverse abdominal respiration, as the name suggests, the lower abdomen retracts during inhalation and expands during exhalation. This form of breathing coordinates the heart and kidneys, as well as the water and fire elements in the body. Routines such as "Shaking the Heavenly Column," "Hugging *Kunlun*," "Rotating Winches," "Propping up the Sky and Pressing the Skull," "Bending to Touch Feet," "Caressing the Belly and Abdomen" and "Rinsing Mouth and Swallowing Saliva" all demand reverse abdominal respiration.

4) Anus-lifting Respiration

This breath form requires consciously tightening the muscles in the anus and groin during inhalation and relaxing the muscles in the anus and groin during exhalation. Anus-lifting respiration can invigorate the kidneys, strengthen *yang*, tone the kidneys to control nocturnal emission, tone *Qi*, clear the *Ren*, *Du* and *Chong* meridian channels, and prevent and cure hemorrhoids and urogenital problems such as prostate disorders. Reverse abdominal respiration in *Shi Er Duan Jin* is used in coordination with anus-lifting respiration.

5) Holding the Breath

Holding the breath in *Shi Er Duan Jin* is done with full lungs. Duration depends on the requirement of each movement and on the exerciser's own conditions. Holding your breath is good for removing turbid *Qi*, thus facilitating air exchange; clearing meridian channels; and improving blood circulation. It can also increase the stimulus intensity and enhance the effects of movements on the joints, muscles, viscera, nerves and body fluids. Rubbing your hands together before Routine 8 "Massaging *Jingmen* on the Back" in *Shi Er Duan Jin* and the first movement in the closing position both require holding your breath.

2. Mind

Mind is an important aspect of *Qigong* practice. Controlling the mind is called "regulating the heart" in *Qigong*. "Heart" regulation means actively and consciously controlling your thoughts to meet the requirements and accomplish the goals of the exercises. Mental activities in *Shi Er Duan Jin,* except for those under special circumstances, mostly involve conscious focus on the standards, key points, key parts and breath requirements of the movements.

Mind application methods in *Shi Er Duan Jin* mainly include: calming the heart, counting silently, concentrating the mind, meditating and thinking.

1) Calming the Heart

Calming the heart means "quieting the thoughts, forgetting both the self and the external world, with the body like a weeping willow and the heart like ice." Calming is good for banishing distracting thoughts, pacifying the mind and *Qi*, and improving the effect and movement of *Qi*. Routine 1 "Calming Heart and Positioning Hands" in *Shi Er Duan Jin* requires this mental discipline.

2) Counting Silently

Counting silently means counting breaths or motion repetitions in the mind. It helps concentrate the mind, enter into meditation, and replace ten thousand thoughts with a single thought. Routine 2

"Tapping Teeth and Sounding Drums" in *Shi Er Duan Jin* requires this mental practice.

3) Concentrating the Mind

Concentrating the mind means focusing on one body part or movement. In *Shi Er Duan Jin,* this includes concentrating on a key movement and concentrating on an acupoint. For instance, "Shaking the Heavenly Column" and "Hugging *Kunlun*" require concentrating on the *Dazhui* acupoint; "Propping up the Sky and Pressing the Skull" and "Bending to Touch Feet" require concentrating on key motions.

4) Meditating and Thinking

Meditating and thinking involves closing the eyes to visualize a certain body part or to imagine a natural landscape. It is a good technique for replacing ten thousand thoughts with a single thought, examining feelings, concentrating the mind and pacifying the heart. In *Shi Er Duan Jin,* Routine 11 "Rotating the Upper Body" requires the exerciser to visualize the *Haidi* acupoint, and Routine 12 "Rinsing Mouth and Swallowing Saliva" requires the exerciser to imagine that the saliva produced in the mouth is swallowed and transported to the *Dantian.*

Section III Basic Postures

1. Sitting Postures

Sitting postures of *Shi Er Duan Jin* include the cross-legged position, half-lotus position and full lotus position. In each posture, the hands lie relaxed on the knees, palms down, thighs should be flat, and leg positions can be alternated between routines.

1) Cross-legged Position

Sit erect. Cross your ankles, with the left ankle closer to the body. Put the feet under the thighs, with the soles facing outward and back (Fig. 6).

Fig. 6

2) Half-lotus Position

Sit erect, with the left heel touching the *Huiyin* acupoint, the right foot upon the left leg and close to the root of the left thigh, with the right sole facing upward (Fig. 7).

Fig. 7

3) Full Lotus Position

Sit erect. Put the right leg upon the left, close to the root of the left thigh, with the right sole facing upward. Then put the left foot up on the right leg, next to the root of the right thigh, with the sole facing upward (Fig. 8).

Fig. 8

2. Points for Attention

1) Warm up the waist and legs before sitting down.

2) Beginners should use the cross-legged position.

3) Half-lotus and lotus positions should follow in proper sequence and be adopted under the guidance of your coach.

4) If your waist or legs hurt or get numb, adjust the position or alternate your leg positions.

3. Cushion

The cushion for *Shi Er Duan Jin* is square, and the side length is about 60 cm. The cushion is about 5 cm thick at the back and 2 cm in the front, to form a slope from back to front. The material should neither be too hard nor too soft.

CHAPTER IV

Descriptions of the Routines

Section I Names of the Movements

Ready Position
Routine 1 Calming Heart and Positioning Hands (*Ming Xin Wo Gu*)
Routine 2 Tapping Teeth and Sounding Drums (*Kou Chi Ming Gu*)
Routine 3 Shaking the Heavenly Column (*Wei Han Tian Zhu*)
Routine 4 Hugging *Kunlun* (*Zhang Bao Kun Lun*)
Routine 5 Rotating Winches (*Yao Zhuan Lu Lu*)
Routine 6 Propping up the Sky and Pressing the Skull (*Tuo Tian An Ding*)
Routine 7 Bending to Touch the Feet (*Fu Shen Pan Zu*)
Routine 8 Massaging *Jingmen* on the Back (*Bei Mo Jing Men*)
Routine 9 Caressing the Belly and Abdomen (*Qian Fu Wan Fu*)
Routine 10 Burning Navel Ring (*Wen Xu Qi Lun*)
Routine 11 Rotating the Upper Body (*Yao Shen Huang Hai*)
Routine 12 Rinsing Mouth and Swallowing Saliva (*Gu Shu Tun Jin*)
Closing Position

Section II Movements, Points for Attention and Functions and Effects

Ready Position

Movements

1. Stand straight, weight centered, with your feet together and arms hanging loosely at your sides. Look straight ahead (Fig. 9).

Fig. 9

2. Slightly bend your right knee. Take a step back with the left foot, with the toes touching the floor, heel lifted. Look straight ahead (Fig. 10).

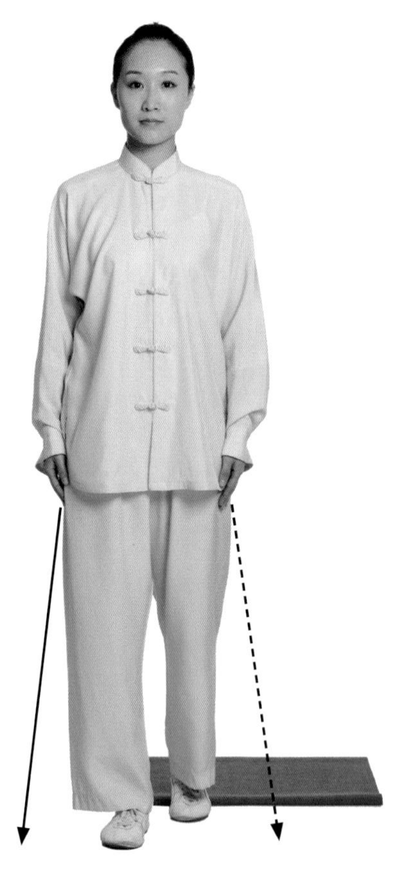

Fig. 10

3. Bend both knees to squat, with the lower right leg erect and with your fingers touching the floor for balance. Slightly bend the elbows and move your body weight forward, looking down and ahead (Fig. 11).

Fig. 11

4. Put the right foot under the left thigh, with the outside of the right calf touching the floor, looking down and ahead (Fig. 12).

Fig. 12

5. Shift your weight to the left and sit straight and cross-legged. Put your hands naturally on top of the knees. Look straight ahead (Fig. 13).

Fig. 13

Points for attention

Maintain an even pace, stay balanced, and sit straight.

Functions and effects

This initial seating routine helps you coordinate your limbs, keep the body straight and centered, regulate your breathing, and calm the mind.

Routine 1 Calming Heart and Positioning Hands (*Ming Xin Wo Gu*)

Movements

1. (Continue from the "Ready Position" routine) Point your fingers upward and raise your arms to form a 45-degree angle between your arms and upper body. Turn the arms outward and raise them further, with the elbows slightly bent. Raise your head, looking up and forward (Fig. 14).

Fig. 14

2. Pull in your chin, turn the arms inward, and bring down to reach directly forward from the shoulders, shoulder-width apart and palms down. Look straight ahead (Fig. 15).

Fig. 15

3. Bring your arms down further. Hold your fingers naturally, with the thumbs touching the base pads of the ring fingers, and put the hands on top of the knees. Lightly close your eyes for 30 seconds (Fig. 16).

Fig. 16

Points for attention

1. Expand your chest and upper body while raising the arms. Keep your neck and spine erect and lift the *Baihui* acupoint while bringing the arms down.

2. Pacify the heart and *Qi*, forgetting the self and the external world.

Functions and effects

1. Calming the heart helps purify the mind, keeps body and mind healthy, coordinates the heart and *Qi*, and arouses the

circulation of *Qi* in the body. Positioning the hands helps to soothe the liver and lungs.

2. This routine can to some extent prevent or cure palpitations, insomnia, giddiness, fatigue and neurasthenia.

Routine 2 Tapping Teeth and Sounding Drums (*Kou Chi Ming Gu*)

Movements

1. (Follow from the previous "Calming" routine) Open your hands and move them up along the waist, turn the arms forward, and hold your arms straight out to the sides at shoulder level. Palms face forward. Look straight ahead (Fig. 17).

Fig. 17

2. (Continue) Bend your elbows and stretch your hands into *Tongtian* Finger; insert your middle fingertips in your ears. Tap your teeth 36 times, looking down and ahead (Fig. 18).

Fig. 18

3. Remove your fingertips from the ears, looking down and ahead (Fig. 19).

Fig. 19

4. Press your palms over your ears, with all ten fingers touching the back of the head and the middle fingers on the bump of the occipital bone. Press the middle fingers with the index fingers and then tap the index fingers on the back of the head 24 times on both sides, looking down and ahead (Fig. 20).

Fig. 20

5. Remove your palms from the ears, and hold your hands in front of your abdomen, palms down. Look straight ahead (Figs. 21–22).

Fig. 21

Fig. 22

Points for attention

1. Push your middle fingers tightly into your ears and count silently while tapping your teeth and the back of your head. Tap your teeth gently, with your lips only slightly closed.

2. Breathe in and out through the nose. Strike the back of your head lightly with the index fingers when "sounding the drums."

Functions and effects

1. Tapping your teeth helps to strengthen them and to prevent or cure dental disorders.

2. "Sounding the drums" helps to refresh the mind and to improve hearing and vision.

Routine 3 Shaking the Heavenly Column (*Wei Han Tian Zhu*)

Movements

1. (Continue directly from the "Tapping and Sounding" routine) Turn your torso to the left about 45 degrees. At the same time, lift the arms and hold them straight out to the sides at shoulder level, with the palms facing back. Fix your gaze on the left hand (Fig. 23).

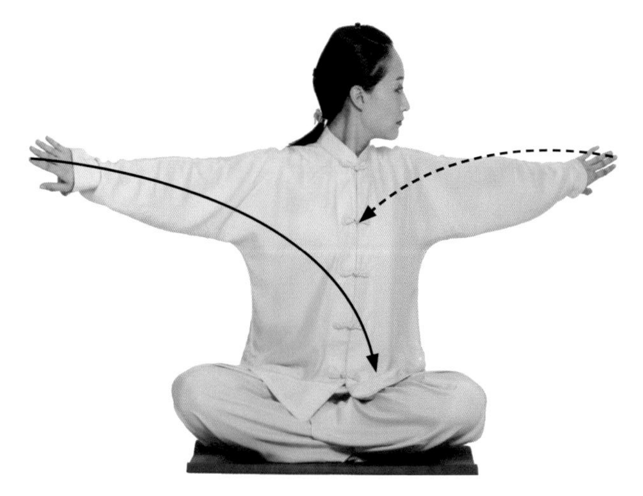

Fig. 23

2. (Continue) Turn the upper body back to center and sit erect. Meanwhile, turn your elbows outward, move your arms forward and hold them in a semicircle in front of the torso, with the left palm above the right facing each other. Look straight ahead (Fig. 24).

Fig. 24

3. (Continue) Move the left hand down to join the right palm in front of the abdomen. Look straight ahead (Fig. 25).

Fig. 25

4. Turn the head to the left, and move both hands together to the inner side of the right thigh, looking to the left (Fig. 26).

Fig. 26

5. Keeping your left shoulder down, push the right palm down with the ball of the left hand. Meanwhile, raise the head and look upward for a short period and then to the left. Hold this position (Fig. 27).

Fig. 27

6. Next, pull in your chin, and turn the torso to the right by about 45 degrees. Then raise your arms and hold them straight out at shoulder level, palms facing back. Fix your gaze on the right hand (Fig. 28).

Fig. 28

Now continue with the sequence above but on the opposite side (Figs. 29–32).

Fig. 29

Fig. 30

Fig. 31

Fig. 32

These movements are to be done three times each, left and right. During the final repetition, pull in your chin and turn your head to face forward. Meanwhile, move your palms slightly to the right, bend your elbows, place your hands on your waist, palms in, thumbs up. Look straight ahead (Fig. 33).

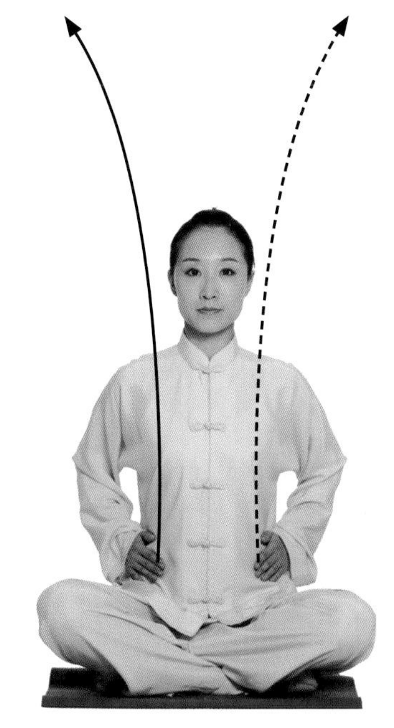

Fig. 33

Points for attention

1. When turning the upper body and arms, initiate the arm motion with the spine, and keep the shoulder down and the spine erect.

2. When turning your head, keep your torso still and neck erect. When raising the head, keep your neck erect without relaxation.

Functions and effects

1. The "heavenly column" refers to the entire cervical spine. Shaking the heavenly column helps stimulate the *Dazhui* acupoint and regulate the *Sanyang* meridians of both the hands and feet, as well as the *Du* channel.

2. Turning the head to the left or right, turning the upper body and arms, and lowering your shoulders help to exercise the spine and to prevent or cure disorders related to the neck, shoulders and waist.

Routine 4 Hugging *Kunlun* (*Zhang Bao Kun Lun*)

Movements

1. (Continue from the "Heavenly Column" routine) Throw your shoulders back and move the arms straight up, with the palms facing each other. Look straight ahead (Fig. 34).

2. Bend the elbows, cross your fingers, and hold the back of your head with your hands. Look straight ahead (Fig. 35).

Fig. 34

Fig. 35

3. Turn the upper body to the left by about 45 degrees, looking to the left (Fig. 36).

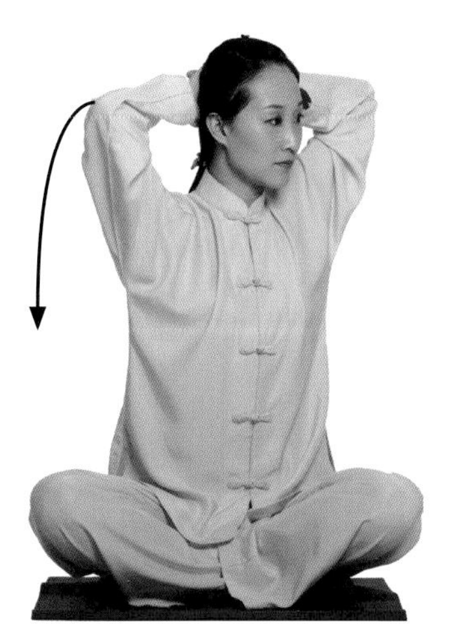

Fig. 36

4. Keep holding your head with your hands. Tilt the torso to the right to stretch the muscles along the left ribs, looking up and to the left (Fig. 37).

Fig. 37

5. Come back upright and sit straight, looking to your left (Fig. 38).

Fig. 38

6. Turn your torso back to center, looking straight ahead (Fig. 39).

Fig. 39

Next, repeat the entire sequence on the opposite side (Figs. 40–43).

Fig. 40

Fig. 41

Fig. 42

Fig. 43

11. Now, raise your head as high as possible, looking up and ahead (Fig. 44).

Fig. 44

12. With your hands still holding the back of your head, put your elbows as close as possible to each other, pull in your chin, and push your head down with your hands. Fix your gaze on your abdomen (Fig. 45).

Fig. 45

13. Separate your hands and move them along your cheeks until the balls of the palms reach your chin. Raise your head and look straight ahead (Fig. 46)

Fig. 46

14. Raise the head, tilt it back and cradle your chin in your palms, looking up and ahead (Fig. 47).

Fig. 47

15. Pull in your chin, and keep your neck erect. Meanwhile, push your hands down in front of the abdomen, turn the elbows outward, and place your palms on your waist, with fingertips pointing across your abdomen. Look straight ahead (Figs. 48–49).

Fig. 48

Fig. 49

These movements are to be done three times each. During the final repetition, after pressing the palms down toward the abdomen, turn your hands into clenched fists and draw them back to the sides of your waist. Look straight ahead (Fig. 50).

Fig. 50

Points for attention

1. When holding your head and turning the upper body, spread your shoulders and elbows backward. When tilting the upper body to the left or right, the appropriate elbow should rise as much as possible to stretch the lateral muscles.

2. When lowering your head, keep the spine erect and pull in your chin. When raising the head, expand the chest and push your waist forward.

Functions and effects

1. The purpose of this routine is to animate functions of *Sanjiao*, the three cavities of the body housing internal organs. Raising your arms helps regulate functions of the spleen and stomach; tilting the upper body helps to stimulate the liver and gallbladder channels, and to disperse stagnant liver *Qi* and promote bile flow.

2. Pressing your head down with the hands helps stimulate the *Du* and urinary bladder channels and the *Beishu* acupoint, and regulates the functions of related organs. Cradling the chin in your hands helps stimulate the *Dazhui* acupoint. According to *Selecting Acupoints According to Channel* (循经考穴编), stimulating the *Dazhui* acupoint helps to combat disorders related to the heart, liver, spleen, lungs and kidneys, and alleviates stress that results from internalizing anger, grief, worry and fear, as well as hectic fever and night sweats.

Routine 5 Rotating Winches (*Yao Zhuan Lu Lu*)

Movements

1. (Continue from "Hugging *Kunlun*" routine) Move your fists behind you to the *Shenshu* acupoint at the back of the waist, with palms facing out. Look straight ahead (Fig. 51).

Fig. 51

2. Turn the upper body to the left by about 45 degrees. Bend the left elbow and raise your arm and fist to shoulder height. Fix your gaze on the left fist (Fig. 52).

Fig. 52

3. Turn the upper body back to the right, then tilt to the left. Bend your left wrist upward, and stretch the left fist forward and to the left at about 45 degrees. Fix your gaze on the left fist (Fig. 53).

Fig. 53

4. Turn your upper body to the left and straighten the torso. Draw the left fist back to the side of your waist, and bend the left wrist, with the palm side facing backward. Fix your gaze on the left fist (Fig. 54).

Fig. 54

Movements 2–4 are to be done six times continuously, a sequence called "rotating the left winch." During the final repetition, turn your upper body to the right to sit erect, and pull the left fist back to the *Shenshu* acupoint at the lower back, with the palm side facing outward. Look straight ahead (Fig. 55).

Fig. 55

5–7. The same as Movements 2–4, but on the opposite side, a sequence called "rotating the right winch" (Figs. 56–58). During the final repetition, turn your torso to the left and sit straight, pulling your right fist to the *Shenshu* acupoint at the lower back, with the palm side facing outward. Look straight ahead (Fig. 59).

Fig. 56 Fig. 57

Fig. 58

Fig. 59

8. Now broaden your shoulders and throw out your chest, hunch your shoulders, then relax the chest and roll your shoulders forward as much as possible. Keep the shoulders down, looking down and ahead (Figs. 60–62). Repeat the routine, rotating your shoulders forward six times. During the final repetition, straighten your torso to sit erect (Fig. 63).

Fig. 60

Fig. 61

Fig. 62

Fig. 63

9. Rotate your shoulders backward six times (Figs. 64–66). During the final repetition, straighten your torso to sit erect (Fig. 67).

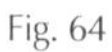

Fig. 64

Fig. 65

Fig. 66

Fig. 67

10. Relax your fists, pointing the fingertips downward. Move the palms upward along the ribs, with the thumbs rubbing the ribs, to rest on the tops of the shoulders, keeping shoulders and elbows down. Look straight ahead (Fig. 68).

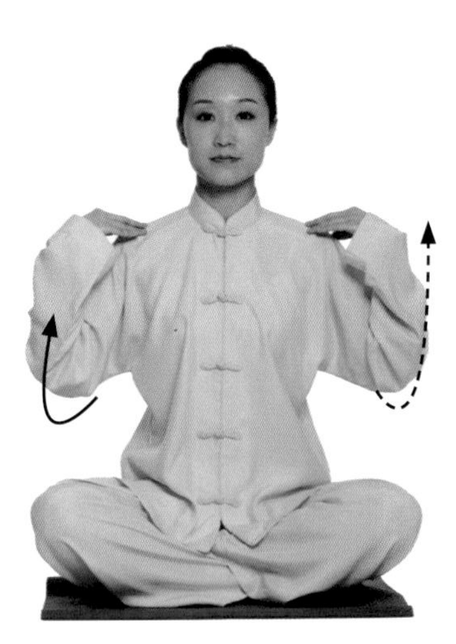

Fig. 68

11. Keeping your hands on top of the shoulders, rotate your torso to the left. Move the right arm forward and the left arm backward, looking down and ahead (Fig. 69).

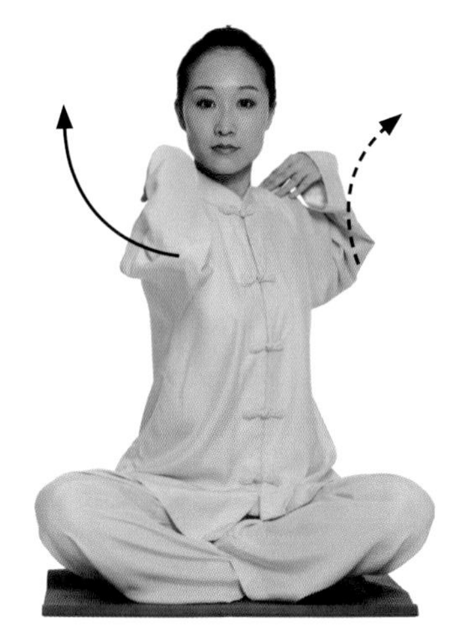

Fig. 69

12. Rotate the torso back, lifting your arms and pointing the elbows up, looking down and ahead, fingertips still on the shoulders (Fig. 70).

Fig. 70

13. Rotate your torso to the right. Move the left arm forward and the right arm backward, looking down and ahead (Fig. 71).

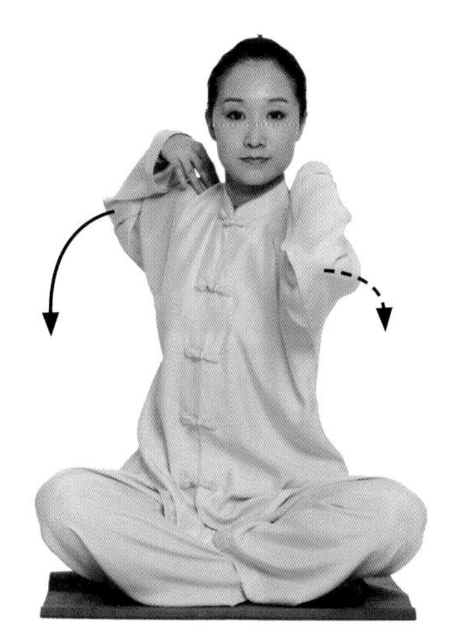

Fig. 71

14. Rotate back, and drop your elbows, fingertips on the shoulders, looking down and ahead (Fig. 72).

Fig. 72

Movements 11–14 are to be done six times, as are Movements 15–18 on the opposite side (Figs. 73–76).

Fig. 73

Fig. 74

Fig. 75

Fig. 76

Points for attention

1. Single rotation (rotating the left or right winch): When stretching the fist forward, turn the waist, lower the shoulder on the same side and bend the wrist upward. When drawing the arm back, bend the elbow and bend the wrist downward.

2. Double rotation (rotating both shoulders forward or backward): Use the base pads of the index fingers to massage the *Shenshu* acupoint, and rotate the shoulders in a flowing manner.

3. Crisscross rotation (alternately moving one shoulder forward and the other backward): Initiate the arm movements with the waist, and move the elbows simultaneously.

Functions and effects

1. This routine helps stimulate the *Sanyin* and *Sanyang* meridians in the hands, the *Du* and urinary bladder channels and the *Beishu* acupoint; it regulates the functions of related organs, clears the heart and lungs, and tones the kidneys and *yang*.

2. The routine helps strengthen the spine and prevents or cures disorders of the shoulders and cervical spine.

Routine 6 Propping up the Sky and Pressing the Skull (*Tuo Tian An Ding*)

Movements

1. (Continue from "Rotating Winches" routine) Lift your elbows to shoulder height, placing the fingertips on your shoulders. Look straight ahead (Fig. 77).

Fig. 77

2. Move the palms down along your ribs, with the thumbs caressing the ribs, to the buttocks, looking down and ahead (Fig. 78).

Fig. 78

3. Turn the arms outward, slide your palms along the outer thighs to the knees, and pull the knees up. Look down and ahead (Fig. 79).

Fig. 79

4. Move your right leg forward, with the toes facing up. Slightly bend the right knee, fixing your gaze on the right foot (Fig. 80).

Fig. 80

5. Move your left leg forward, and straighten both legs, with the toes facing upward. Lay your hands lightly on top of your knees, looking at your toes (Fig. 81).

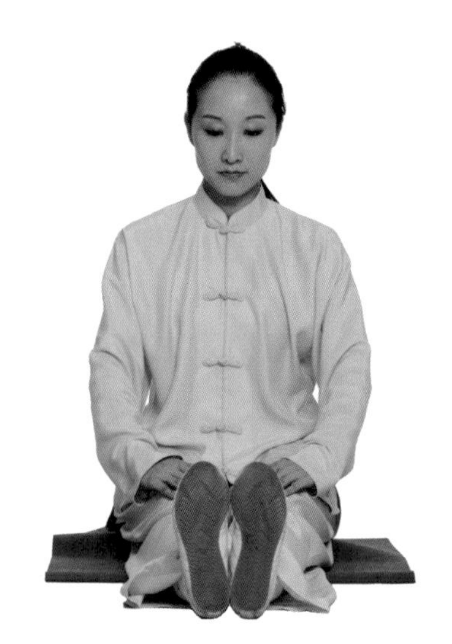

Fig. 81

6. Turn your arms outward, and elevate your palms in front of the abdomen, with the fingertips pointing toward each other and both palms up. Interlock your fingers, looking down and ahead (Fig. 82).

Fig. 82

7. Raise your hands up in front of your chest, turn the arms inward, keep your palms joined and straighten your arms. Meanwhile, keep the knees straight, and push your feet forward, looking down and ahead (Fig. 83).

Fig. 83

8. Keep the shoulders down, bend your elbows, turn your hands 180° and put them on top of your head, slightly pressing down. Reverse your foot stretch, with toes now pointed toward your shins, looking down and ahead (Fig. 84).

Fig. 84

9. Turn the arms inward, join your palms again and straighten your arms upward as in Fig. 83. Meanwhile, keep your knees straight, stretch your feet forward, and look down and ahead (Fig. 85).

Fig. 85

These movements are to be done nine times, with the hands raised and pressed down on the head once each time. The final movement during the last repetition is the same as 8 above (Fig. 86).

Fig. 86

Points for attention

1. When the hands are raised, keep your torso and arms straight, stretch the arms and the spine from the waist, stretch the muscles from the ribcage, keep your knees straight, and stretch the feet.

2. When pressing down on your head, keep your lower back erect, your neck craned to push the head up, knees straight, and toes pointed toward your shins.

Functions and effects

1. Stretching your feet back and forth helps to stimulate the *Sanyin* and *Sanyang* meridians in the feet, clears the channels and collaterals, and facilitates the circulation of blood and *Qi* in your body.

2. Stretching the spine, ribcage muscles, shoulders and neck helps to animate the functions of *Sanjiao*, soothe the liver and gallbladder, and prevent or cure disorders of the shoulders and neck.

Routine 7 Bending to Touch the Feet (*Fu Shen Pan Zu*)

Movements

1. (Continue from the "Propping and Pressing" routine) Separate your hands and hold your arms straight up, palms facing each other. Relax the ankles, with your toes pointing upward. Look straight ahead (Fig. 87).

Fig. 87

2. Bend forward by no more than 45 degrees, and reach out with the hands to grasp the soles of your feet, with the thumbs pressing the insteps. Fix your gaze on your toes (Fig. 88).

Fig. 88

3. Pull your feet to keep the toes pointed back toward your shins. Keep the knees straight, push the waist forward, raise your head and hold this position. Look upward (Fig. 89).

Fig. 89

4. Keeping your legs straight and lower back bent, pull in your chin and stretch the muscles of the neck. Hold this position. Fix your gaze on the knees (Fig. 90).

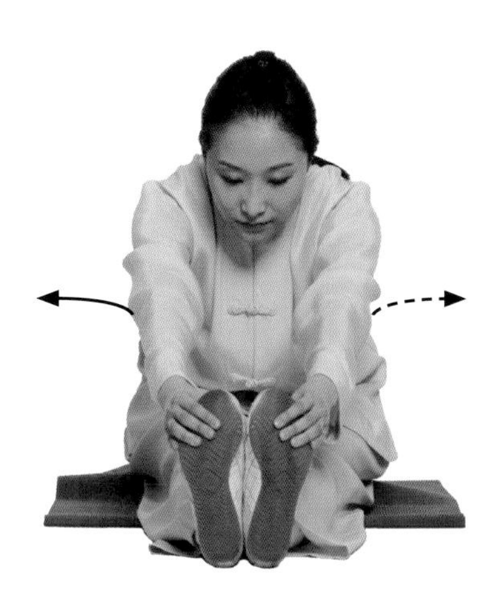

Fig. 90

5. Straighten your torso and neck, and release your feet, with palms down. Bend your elbows, draw your hands back along the legs to the waist, straighten the arms and push your hands backward, with the palms facing back. Look straight ahead (Fig. 91).

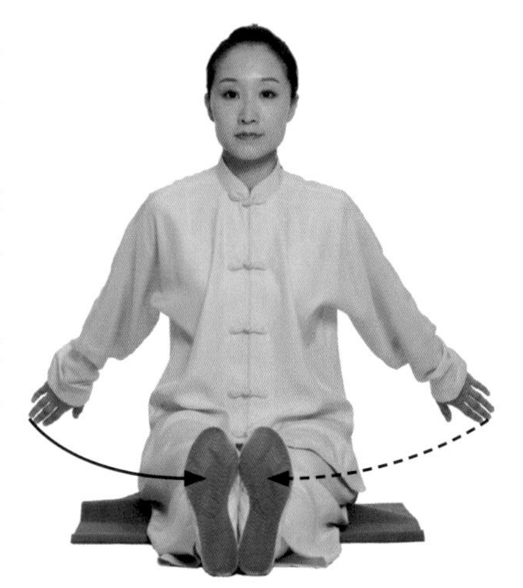

Fig. 91

6. Bend forward by no more than 45 degrees, turn your elbows outward, and move the hands forward to grasp the feet, with your thumbs pressing the insteps. Fix your gaze on the toes (Fig. 92).

Fig. 92

7–8. Repeat Movements 3–4.

Repeat Movements 5–8 four times, which means Movements 3–4 and 7–8 are done six times altogether. During the final repetition, straighten the upper body and neck, release your feet, lay your hands palm down on your knees, and look down and ahead (Fig. 93).

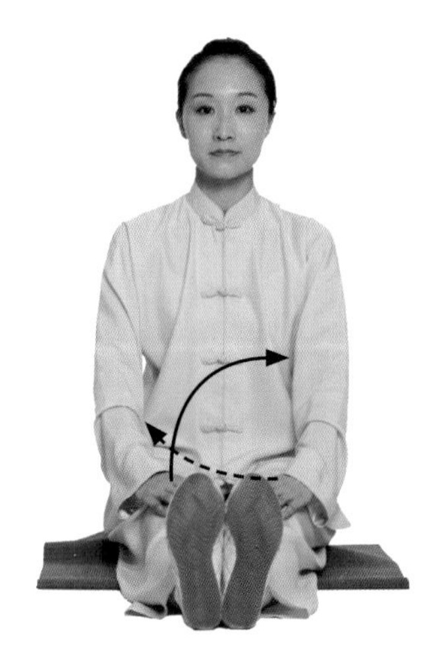

Fig. 93

9. Turn the left elbow outward, with the palm up, and move the hand in a curve to the right. Move the right arm up above the left arm, with the palm down, fingers pointing up, and curve the right palm up to the left. The forearms should be parallel to each other in front of the abdomen. Fix your gaze on the right palm (Fig. 94).

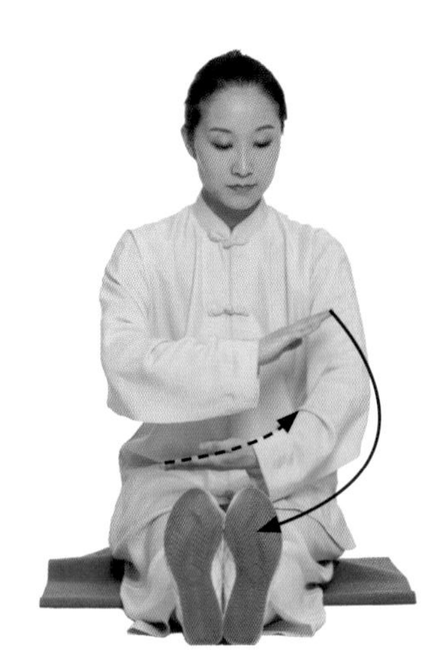

Fig. 94

10. Turn the left elbow inward, and press the left palm to the root of the left thigh. Bend forward, and reach with the right hand to grasp and pull the left sole, looking at the left foot (Fig. 95).

Fig. 95

11. Now straighten the torso, and slightly bend the right knee. Bend the left leg, and place it beneath the right thigh with your right hand, looking downward (Fig. 96).

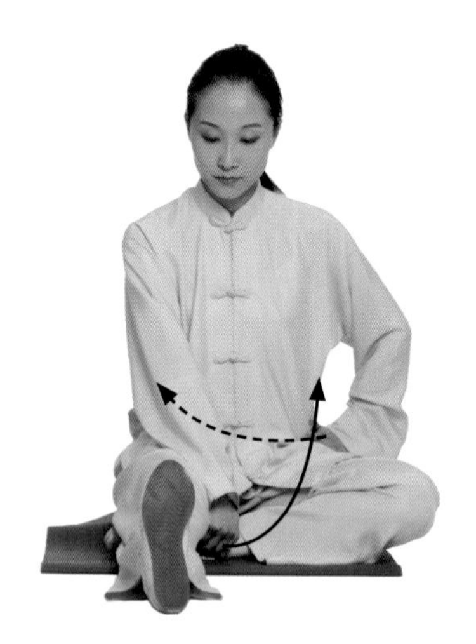

Fig. 96

12. Turn the right elbow outward, and with palm up, move the hand in a curve to the left. Move the left arm across the chest and above the right arm, palm down, and move the left hand to the right. The arms parallel each other in front of the body. Fix your gaze on the left palm (Fig. 97).

Fig. 97

13. Turn the right elbow inward, and press the right palm to the root of the right thigh. Bend forward, and reach with the left hand to grasp and pull the right sole, looking at the right foot (Fig. 98).

Fig. 98

14. Straighten the torso, and slightly bend the left knee. Bend the right leg, and place it beneath the left thigh with your left hand, looking downward to the left (Fig. 99).

Fig. 99

15. Straighten your upper body, and press your left palm to the root of the left thigh, looking down and ahead (Fig. 100).

Fig. 100

Points for attention

When expanding your chest, push the waist forward and straighten the knees, keeping the toes pointed toward the shins. When raising your head, move the chin upward. When pulling in the chin, crane your neck.

Functions and effects

1. This routine helps to stimulate the *Ren*, *Du*, and Belt channels, other channels and collaterals, and exercises the spine, neck and muscles of the waist and back.

2. Modern medicine holds that exercising the spine at the waist helps stimulate spinal and vegetative nerves, and can to some extent cure brain disorders and boost intellectual resources; straightening the legs and stretching the toes back toward the shins helps to stretch the *cauda equina* nerve and eases muscular pain.

Routine 8 Massaging *Jingmen* on the Back (*Bei Mo Jing Men*)

Movements

1. (Continue from "Bending to Touch Feet" routine) Bend forward, and extend your arms backward, with the palms up. Look down and ahead (Fig. 101).

Fig. 101

2. Move your arms laterally to the sides, with the palms facing up. Look down and ahead (Fig. 102).

Fig. 102

3. Straighten the torso, extend the arms directly forward, to form a right angle with your upper body, with the arms shoulder-width apart and the palms facing down. Look straight ahead (Fig. 103).

Fig. 103

4. Bend the elbows and put your palms together before the chest, with fingertips pointing upward. Look down and ahead (Fig. 104).

Fig. 104

5. Press your palms tightly together, with the left palm on top of the right., placing the fingertips of one palm at the ball of the other; level the hands and move them in front of the abdomen. Look down and ahead (Fig. 105).

Fig. 105

6. With the palms tightly together, raise them a little, and turn the wrists to put the right palm on top of the left; level the hands again and move them in front of the abdomen. Look down and ahead (Fig. 106).

Repeat seven times. During the final repetition, place the left palm on top of the right.

Fig. 106

7. Bend your elbows and move your hands back along the sides of the abdomen to the lower back, and point your fingertips downward. Look down and ahead (Figs. 107 and 107 back).

Fig. 107

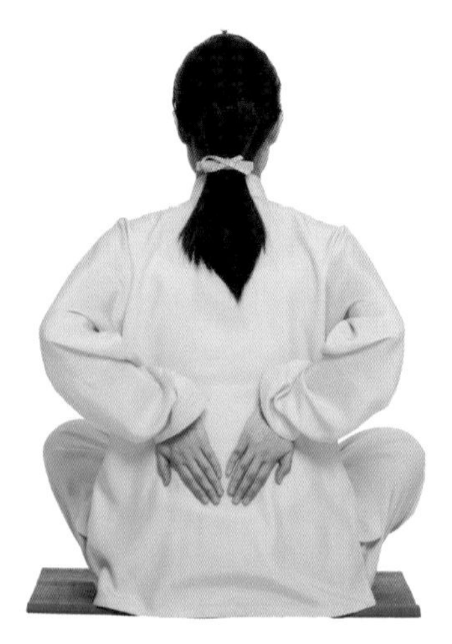

Fig. 107 back

8. Move your hands up and down to massage the back of the waist with the palms. Look down and ahead. Repeat 24 times, with the palms moving up and down once each time.

Points for attention

1. When rubbing your hands, hold your breath, press your palms tightly against each other and rub them till they become warm.

2. When massaging the lower back, keep your fingers together and the hand in a hollow cup position. Apply more force when massaging downward and less when massaging upward, using a moderate speed.

Functions and effects

Jingmen, a *Qigong* term, finds its origin in *Ten Works on Cultivating Perfection: Zhong Li Ba Duan Jin Fa* (修真十书・钟离八段锦法): "*Jingmen* is above the kidneys at the back of the waist."

Massaging the *Shenshu* and *Yaoyan* acupoints helps to warm and clear related channels and collaterals, tone the kidneys and *Qi*, and effectively prevent or cure backache, leg weakness, impotence and dysmenorrhea.

Routine 9 Caressing the Belly and Abdomen (*Qian Fu Wan Fu*)

Movements

1. (Continue from the "Massaging *Jingmen*" routine) Place, and move your palms along the ribs to the lower chest, with the fingertips pointing toward each other. Look down and ahead (Fig. 108).

Fig. 108

2. Turn the fingertips downward and move your palms down along the front of the abdomen. Look down and ahead (Fig. 109).

Fig. 109

3. Move your hands to the sides of your abdomen, with fingertips pointing down and toward each other and the hands tilting outward. Look down and ahead (Fig. 110).

Fig. 110

4. With fingertips still pointing downward, move your hands up along the ribs to your lower chest, with your fingertips pointing to each other. Look down and ahead (Fig. 111).

Fig. 111

These movements are to be done six times, with the hands moving down and up once each time. During the final repetition, move the palms further down along the front of the abdomen, with your fingertips pointing downward. Look down and ahead (Fig. 112).

Then move your hands up and repeat again six times (Figs. 113–115).

Fig. 112

Fig. 113

Fig. 114

Fig. 115

During the final repetition, put your palms on the ribs, with fingertips pointing to each other (Fig. 116).

Fig. 116

Points for attention

1. When caressing the belly from bottom to top, inhale, pull in your buttocks and contract your abdomen. When caressing the belly from top to bottom, exhale and relax your buttocks and abdomen.

2. This routine should be practiced at an even pace and with reasonable force.

Functions and effects

1. Massaging the abdomen helps regulate the circulation of blood and *Qi*, clears channels and collaterals, and facilitates the circulation of blood in the internal organs in the abdominal cavity.

2. This routine helps disperse stagnant liver *Qi*, regulates *Qi* and the internal organs, and improves the functions of the digestive and urogenital systems.

Routine 10 Burning Navel Ring (*Wen Xu Qi Lun*)

Movements

1. (Continue from the "Caressing" routine) Overlap your hands at the navel, with the left palm touching the navel. Lightly close your eyes, and concentrate the mind on the navel for 2 to 5 minutes (Fig. 117).

Fig. 117

2. Open your eyes, gently massage around the navel ring with the palms clockwise in three circles, and then massage counterclockwise in three circles. Look down and ahead.

Points for attention

1. Imagine that the navel ring warms up, begin obverse abdominal respiration, and keep the body erect, centered and in a comfortable position.

2. When massaging the navel ring, point the *Laogong* acupoint at the center of the palms to the navel, massage slowly and gently, and breathe naturally.

Functions and effects

Chinese medical theories and traditional *Qigong* maintain that the navel is the root of human life and an important acupoint in the *Ren* channel.

1. Concentrating on the navel ring helps to tone *Qi*, soothe the nerves, enhance circulation and vitality, improve coordination between the heart and kidneys, and regulate the balance of *yin* and *yang*.

2. This routine helps relax the cerebral cortex neural cells, regulate brain activities, improve the efficiency of brain cells, and brain function. It also helps ease the sympathetic nervous system and improve your mood.

3. Massaging the navel ring helps to clear channels and collaterals, regulate the circulation of blood and *Qi*, and prevent the stagnation of *Qi* caused by excessive concentration of the mind.

Routine 11 Rotating the Upper Body (*Yao Shen Huang Hai*)

Movements

1. (Continue from the "Burning Navel" routine) Put your hands on top of the knees. Look straight ahead (Fig. 118).

Fig. 118

2. Lightly close your eyes, tilt the upper body to the left and rotate clockwise from the waist six times. After the final rotation, return to sit erect (Figs. 119–123).

Fig. 119

Fig. 120

Fig. 121

Fig. 122

Fig. 123

3. Tilt the upper body to the right and rotate counter clockwise from the waist six times. After the final rotation, return to sit erect. Open your eyes and look straight ahead (Figs. 124–128).

Fig. 124

Fig. 125

Fig. 126

Fig. 127

Fig. 128

Points for attention

1. When rotating the upper body, keep the spine erect, pull in your chin, and rotate smoothly and consistently at an even pace.

2. Do not rotate excessively or lift your knees.

3. Visualize the *Haidi* acupoint, resettle your inner energy circulation and tune the body.

Functions and effects

The *Huiyin* acupoint, also called *Haidi,* is located between the external genitals and the anus. The *Yinqiao* channel is located here below the scrotum in males, in front of the Weilü (or *Changqiang*) acupoint. Taoists maintain that the channel is the intersection of the *Ren* and *Du* channels, so they start to collect *Qi* from this channel. As long as this channel is clear, other channels will also be clear.

1. Visualizing the *Haidi* acupoint helps to clear the *Ren* and *Du* channels, regulate the circulation of blood and *Qi,* resettle inner energy circulation and tune your body.

2. Rotating the spine helps strengthen the spine at the waist, and massage, stimulate and improve the functions of the abdominal organs.

Routine 12 Rinsing Mouth and Swallowing Saliva (*Gu Shu Tun Jin*)

Movements

1. (Continue from the "Rotating" routine) Extend your arms from the sides of the waist and move them back in an upward arc, palms facing backward. Look down and ahead (Fig. 129).

Fig. 129

2. Turn the elbows out, and hold your arms in a semicircle in front of the abdomen, level with the navel, with the fingertips pointing toward each other. Look down and ahead (Fig. 130).

Fig. 130

3. Bend your elbows and lower the hands. Clench your fists when passing by the navel and place your hands on the roots of the thighs, thumbs up. Look down and ahead (Fig. 131).

Fig. 131

4. Lightly close your lips. Use the tongue to scour the mouth from the right to the left, then to the top and the bottom. Move the tongue outside the teeth, and scour the gums from right to left, then top and bottom. This movement is to be done six times, with the tongue moving inside and outside the teeth once each time.

5. This movement is the same as Movement 4, but is done in a circle, also six times, with the tongue scouring inside and outside of the teeth once each time.

6. Rinse the left and right cheeks with saliva 36 times (also called "drum rinsing"). Don't swallow. Look down and ahead.

7. Turn the elbows outward, relax your fists into natural hands and move them up in front of your chest. Look down and ahead (Fig. 132).

Fig. 132

8. Open and raise your arms straight up, palms facing out. Look straight ahead (Fig. 133).

Fig. 133

9. Turn your elbows outward and relax them slightly, and clench your fists, with the palm sides of the fists facing each other. Look down and ahead (Fig. 134).

Fig. 134

10. Move the fists down and put them on the roots of the thighs, with thumbs up. While moving your fists down, divide the saliva in the mouth into three portions, swallow the first part and imagine the saliva is transported to the *Dantian*. Look down and ahead (Fig. 135).

Fig. 135

Movements 7–10 are to be done three times to swallow all the saliva in three portions.

Points for attention

1. Imagine the mouth is filled up with saliva.
2. The tongue scours the mouth smoothly and consistently.
3. When rinsing the mouth, the cheeks vibrate quickly.
4. Give out a gurgling sound while swallowing the saliva and imagine the saliva is transported to the *Dantian*.

Functions and effects

The saliva, called "divine water," is the most valuable treasure in the world, say Taoists, and the finest subtle matter of the five elements. Ancient Chinese created the character "to live (活)" from "water (水)" and "tongue (舌)" to indicate that water on the tongue is necessary for life.

1. Using the tongue to scour and rinse the mouth helps increase salivary secretion. Saliva can wash and clean the mouth and prevent or cure gingivitis and gingival atrophy.
2. Swallowing the saliva helps to regulate *Qi*, tone the internal organs, nourish the whole body, help digestion, improve circulation, allay fatigue, delay senility and promote health.

Closing Position

Movements

1. (Continue from the "Swallowing" routine) Draw your fists back to the sides of the waist, inhale, roll back the shoulders and expand your chest, holding your breath for two seconds. Stretch your arms forward and cross the wrists in front of your chest, with the left arm nearest the chest and the palm sides of the fists facing back. Apply a little forward force on the fists. Relax your chest, lean the torso back, and pause in position. Look straight ahead (Fig. 136).

2. Relax your fists into palms and place them on top of the knees, palms up. Look straight ahead (Fig. 137).

Fig. 136

Fig. 137

3. Raise your hands high and to the sides at 45-degree angles, with the palms up and the elbows slightly bent. Raise your head and look up and ahead (Fig. 138).

Fig. 138

4. Pull in the chin, turn the arms inwards, lower them halfway and extend your arms and hands directly in front, at shoulder level with the arms shoulder-width apart and palms down. Look straight ahead (Fig. 139).

Fig. 139

5. Place your hands on top of the knees, palm down, and hold this position. Look straight ahead (Fig. 140).

6. Move your palms along your outer thighs, with the fingers touching the floor for balance. Look down and ahead (Fig. 141).

Fig. 140

Fig. 141

7. Bend to move your weight forward. Balance your body with the fingers and feet, and stand up. Look down and ahead (Fig. 142).

Fig. 142

8. Stand erect, and step the right foot forward and ahead of the left. Let your arms hang loose, and look straight ahead (Fig. 143).

9. Move the left foot to the right to bring your feet together. Look straight ahead (Fig. 144).

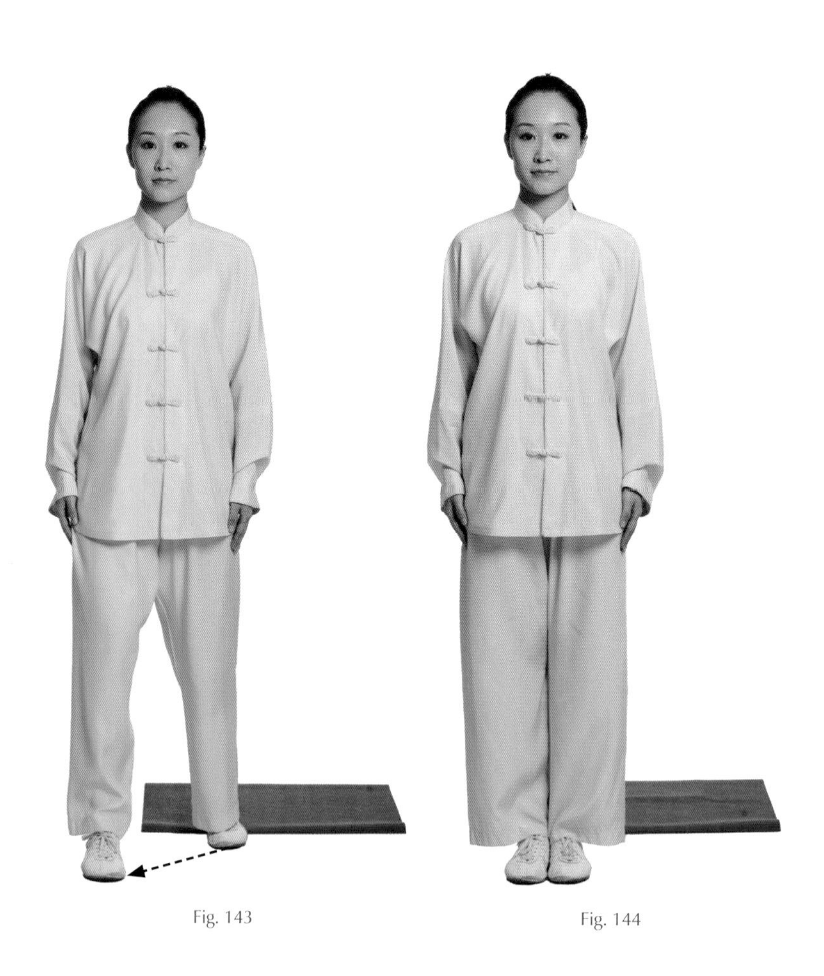

Fig. 143

Fig. 144

Points for attention

1. When crossing your wrists, holding your breath and leaning back, clench the fists tightly, pull in the buttocks, contract the abdomen and clench your teeth. When moving the hands down, imagine your whole body relaxed, with blood and *Qi* circulating well.

2. When holding your hands high, regulate your breathing. When moving your hands down, resettle the inner energy circulation and tune the body.

3. When getting up, use your hands and feet to support yourself, get your weight in balance, and finish the movements in a consistent and steady manner.

Functions and effects

This routine helps to relax the limbs, regulate your breathing, refresh the mind, and regain the normal state experienced before the exercises.

Acupuncture Points

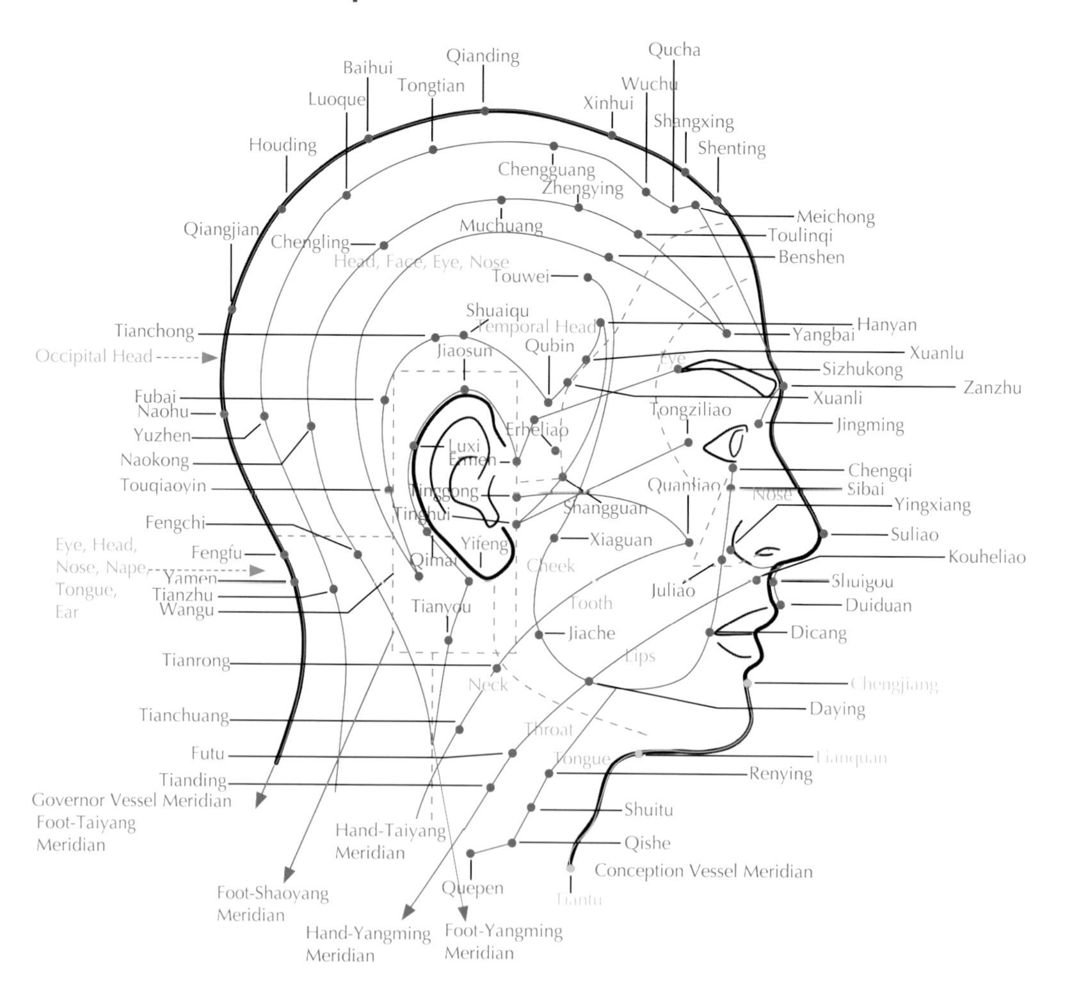

Acupoints on the head and face

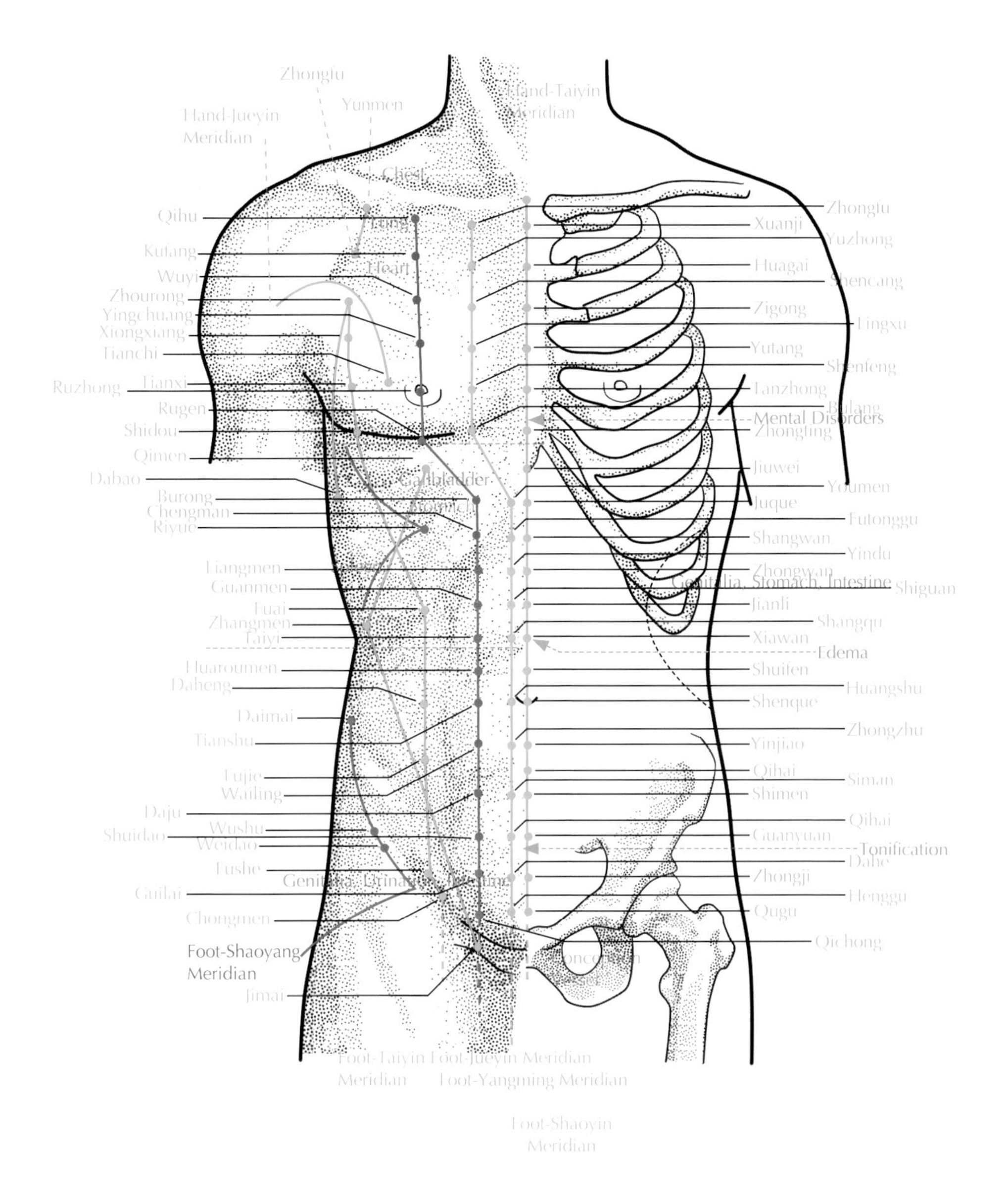

Acupoints on the chest and abdomen

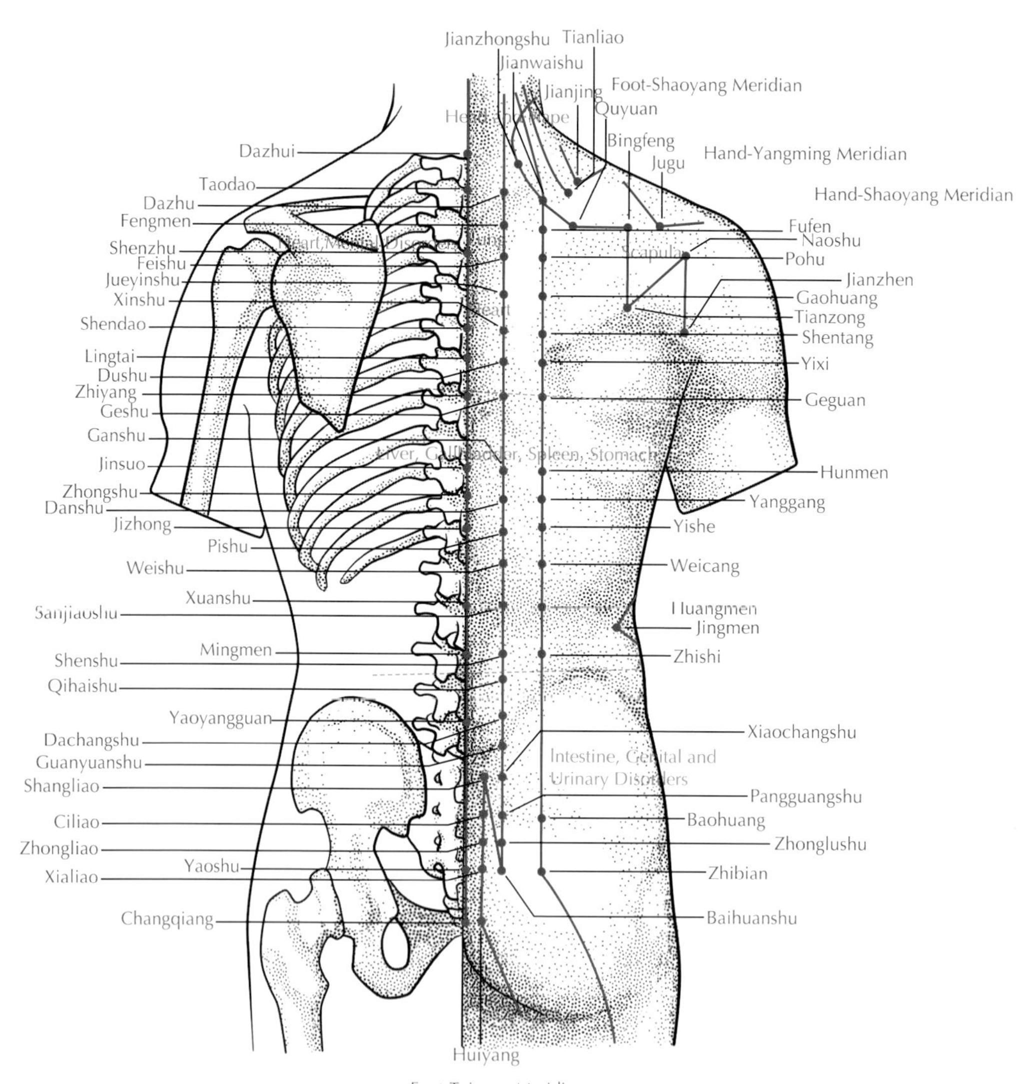

Acupoints on the back and lumbar region

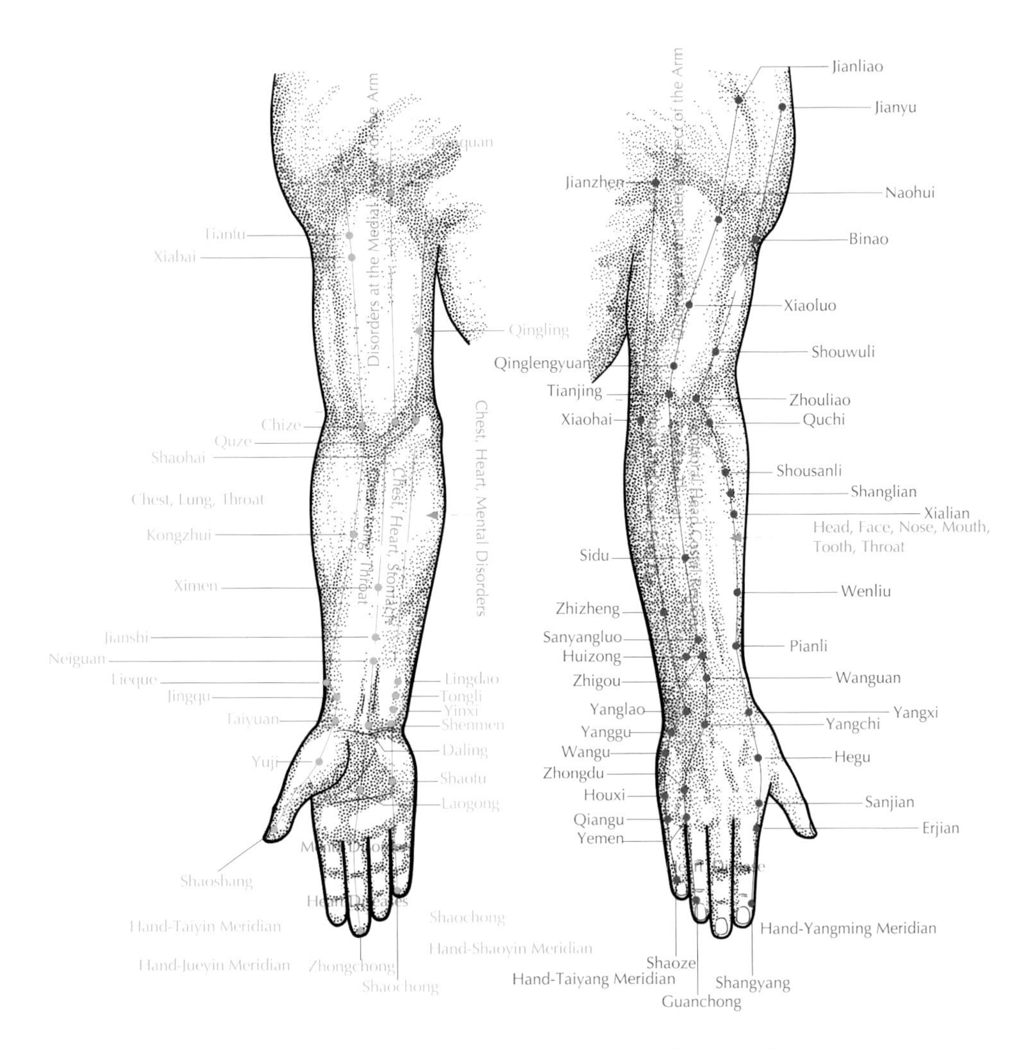

Acupoints in the upper limbs

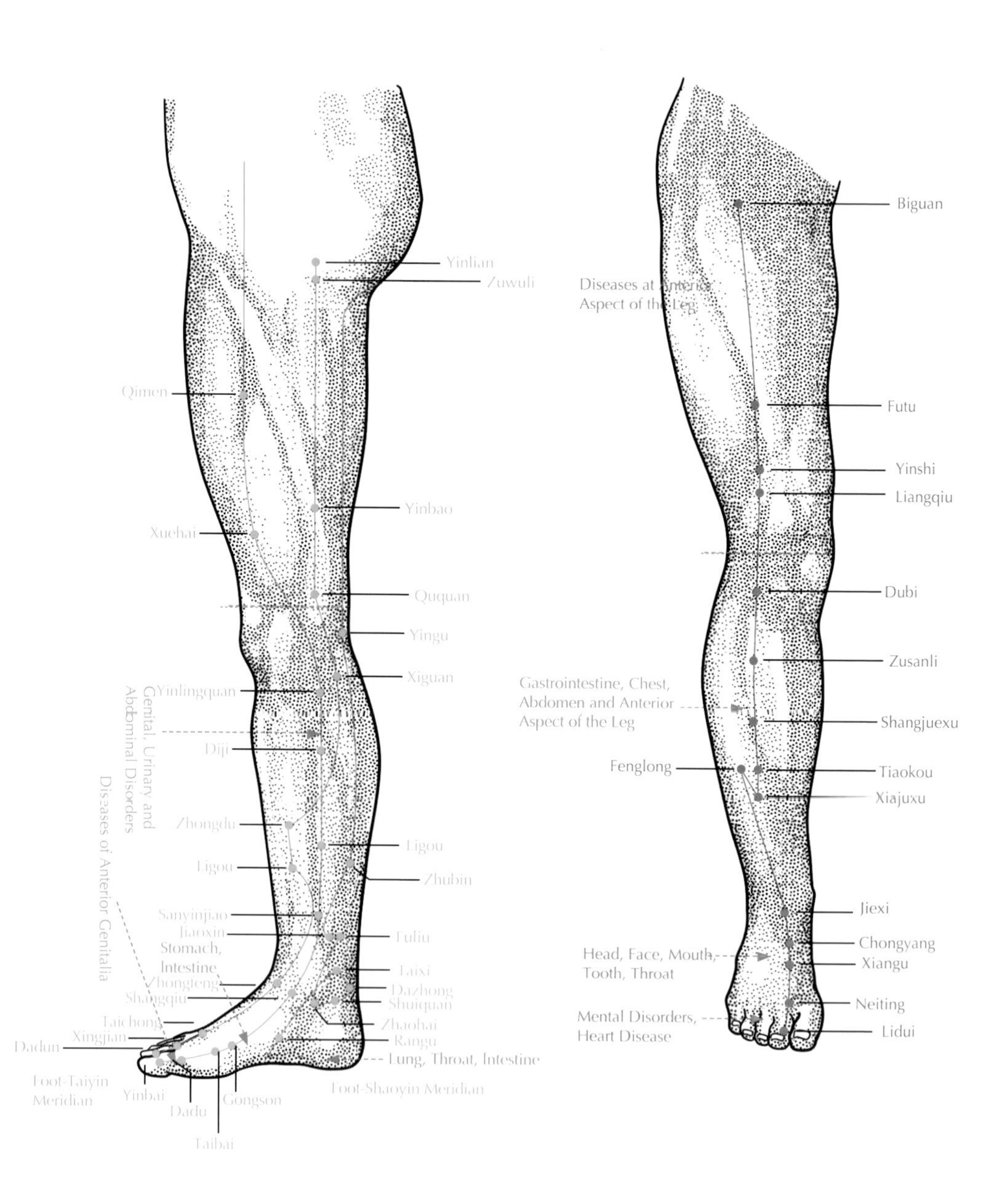

Acupoints in the lower limbs

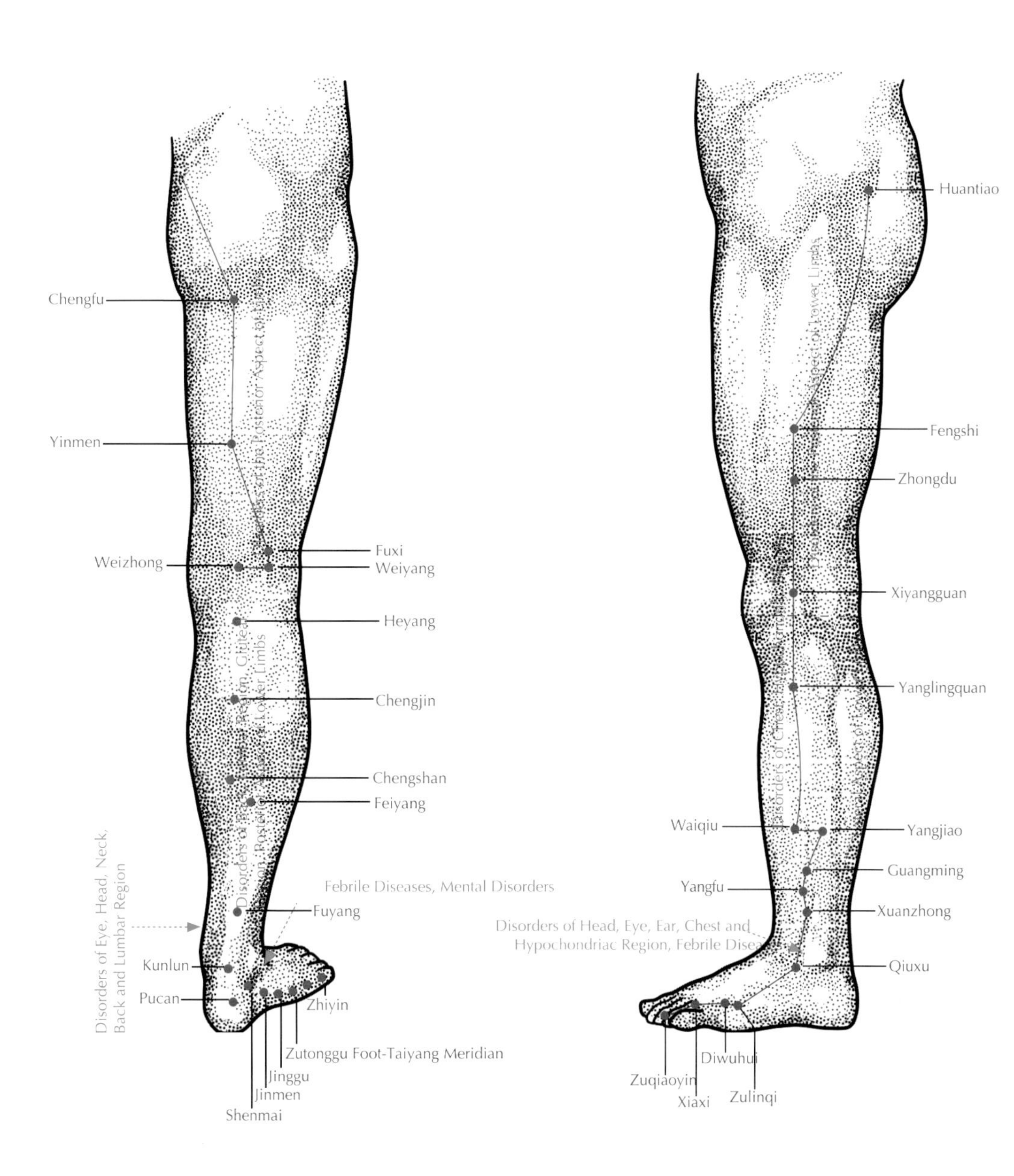

Acupoints in the lower limbs